# DESIGN MANAGEMENT FRAMEWORK

IVY M. A. ABU
THEOPHILUS ADJEI-KUMI

*AuthorHouse™ UK*
*1663 Liberty Drive*
*Bloomington, IN 47403 USA*
*www.authorhouse.co.uk*
*Phone: 0800.197.4150*



*Published by AuthorHouse 01/24/2019*

*ISBN: 978-1-7283-8174-9 (sc)*
*ISBN: 978-1-7283-8175-6 (e)*

*Print information available on the last page.*

*This book is printed on acid-free paper.*



authorHOUSE®

# CONTENTS

# INTRODUCTION

Coordination of design tasks to ensure appropriate quality of information and project delivery within a specified timescale and cost bracket is of paramount interest to all stakeholders in the construction industry.

Building services installation works (BSIW) is a specialized component of a typical building that converts the built environment into one that is comfortable and secure. This is achieved through the creation of conducive internal environments, protection from the external conditions, and the provision of welfare and security. Studies have shown that in Ghana, costs for building services range from 15 to 45 per cent of the total construction costs of a typical building. This range is dependent on the intended use of the building.

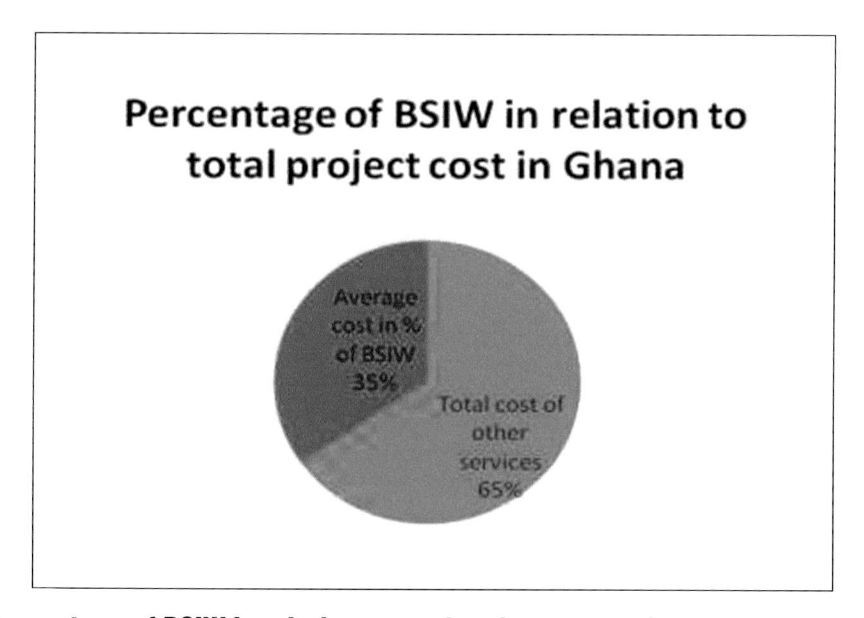


**Figure I.1 Percentage of BSIW in relation to total project cost in Ghana field survey, June 2007)**

# Chapter 1

# The Concept of Design Management

Design management offers organizations an approach to making design-relevant decisions in a market- and customer-oriented way as well as an approach to optimizing design-relevant processes. It is a long, continuous, comprehensive activity carried out at all levels of business. Design management acts in the interface of management and design and functions as a link between the platforms of technology, design, management, and marketing at internal and external interfaces of the enterprise. Figure 1.1 summarizes design management and presents it graphically.

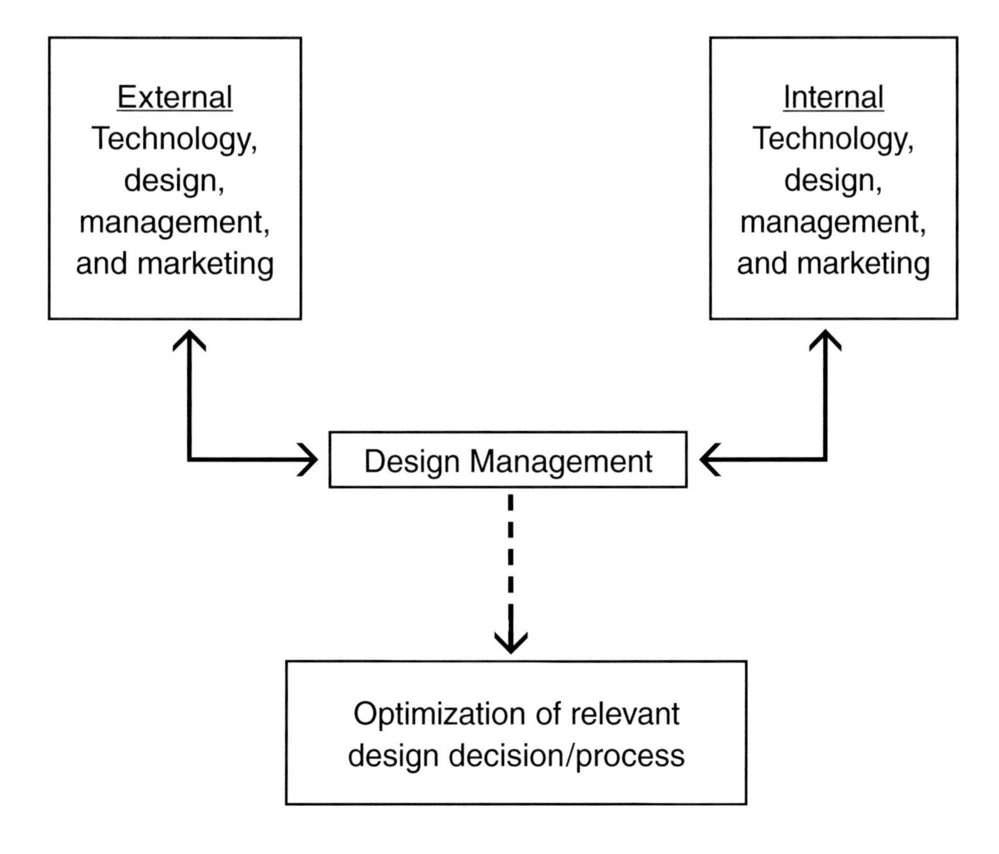


**Figure 1.1 Concept of design management**

# Chapter 2

# Development of a Design Management Framework

The development of an effective design management process encompasses communication and collaboration, as indicated by Gray and Hughes (2006). Irrespective of the form or type of communication, whether among professionals in a particular discipline or those of another, there is a need for consistency in order to achieve effectiveness.

## Development of the Framework

Key factors in addition to effective communication are considered to be integral to achieving the development of a sound framework. These are:

- **cos**t, the monetary value or price of a project or an activity, including the monetary value of all resources required to perform and complete the activity;
- **time**-, the planned dates and durations for undertaking activities as well as meeting milestones;
- **quality**, the degree to which a set of inherent characteristics fulfils a project's requirements;
- **communication**, the process of exchanging information among persons through the use of common systems (e.g. drawings, signs, or symbols);
- **procurement strategy**, the process of acquiring something from inception until the final project is completed;
- **team collaboration** the coordination and cooperation of members of the design team and the construction team as factors to be managed effectively and collaboratively for effective design management.

The development of this framework focuses on these key areas for an operationally, economically, and technically feasible model for Ghana.

## Stages of Construction Considered for the Framework Development for Design Management

Before actual construction starts, a typical design will undergo these clearly defined stages of development

- feasibility studies,
- concept/outline design,
- scheme design,
- detail design,
- tendering

The model development considered all stages, with design management processes occurring in the outline, scheme, and detail stages. A new stage – the initial stage – has been added for the purpose of this framework. This stage is to allow for the client's brief and for the preparation of defined roles and responsibilities before the start of the project. It must, however, be noted that although design management was considered within three stages, the existing stages ought to be integrated, since they all need to exist for an effective design process. Hence, the framework touches on all stages identified.

The framework also describes the relationship among team members in each of these stages, also indicating which member has the power of authority for each particular stage.

## Stakeholders Considered for Effective Design Management

The development of the model considered all professionals in the construction industry, as well as any other individual who would be affected directly or indirectly by a proposed project, such as a stakeholder for the management of the design for that particular project. Since projects carried out in Ghana are not all stereotypes, room is being created for additional stakeholders within the framework, should the need for any arise. Some permanent members, however, included architects, project managers, services designers, and consultants, as well as the client.

It must be well noted that, from studies, stakeholders have the ability to influence major characteristics of a project. Their ability to influence these characteristics is highest at the start of the project and progressively lowers as the project unfolds (PMI, 2008). Figure 2.1 shows a relationship between the influence of a stakeholder and the project timescale.

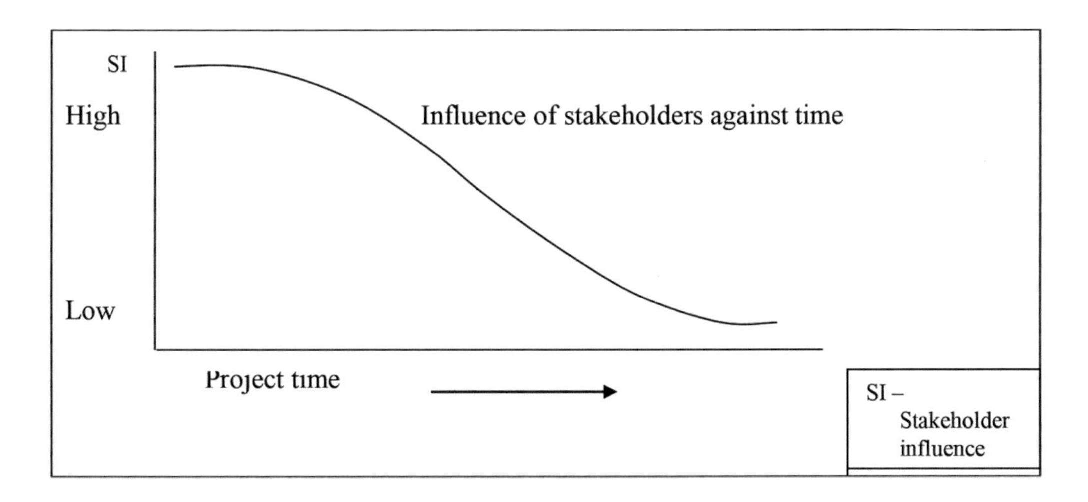


Figure 2.1 **Stakeholders' influence over time** (PMI, 2008).

## Assumptions Made in the Creation of the Framework

The following assumptions were considered in the development of the model framework.

- Building services (BSIW) form a part of the entire umbrella of any typical building project.
- The team leader is part of the project management structure and is the primary link between the entire project's design management
- The design team leader is responsible for coordinating the activities of a design team that may consist of one or more design units with the capacity to cover the scope of the project.

## Application of Design Management Techniques in the Framework Development

Four management techniques were combined for the development of the framework. Table 2.1 presents a summary of all the techniques and their areas of application in the framework.

Design management framework for building services installation works in Ghana

Table 2.1 **Design Management techniques used in development of the framework**

| No. | Design Management technique | Application in developed framework |
|---|---|---|
| 1 | Work Breakdown Structure (WBS) | This technique was used in the outline stage in the development of the framework. It indicates management activities for proper definition and decomposition of design scope. It also informs the cost and schedule plans. |
| 2 | Dependency Structure Matrix (DSM) | This technique was used during the outline stage to indicate design processes based on optimal flow of information. It details what piece of information is needed to start a particular activity and where information or output generated by the activity will lead or become useful. |
| 3 | Value Management | This technique was used during the scheme-design stage. It helps in the revision of alternative designs and aids in the selection of an appropriate scheme that best fits clients' requirements at a lower cost without compromising quality. |
| 4 | Quality Function Deployment (QFD) | This technique was used during the scheme-design stage. It helps in the revision of alternative designs and aids in the selection of an appropriate scheme that best fits clients' requirements at a lower cost without compromising quality. |

# Chapter 3

# Design Management Framework

### Overview

Creating the design management framework requires a consideration of what procurement systems will be in use for carrying out building projects on the whole. The systems as earlier outlined could be traditional, integrated, or management-oriented systems of procurement. Dansoh (2005) indicates that most project management practised in Ghana actually lacks the fundamentals of project management. Osei-Tutu (2000) also indicates in his study that the most practised systems of procurement in Ghana are the traditional and design-and-build strategies. Based on these facts and considering the scope of use for the framework being developed, in regard to Ghana, the researcher considered the application for use for the developed framework with two procurement systems: traditional, and design and build. However, very brief modifications can be made to the model to cater for the management-oriented systems of procurement, if need be. The framework first was intended to develop a management methodology for the entire building process and then was narrowed further to a specific methodology for BSIW. This was based on earlier assumptions that building services always occur in the context of a typical building project. Assignment guides were created in addition to the BSIW framework. They are to serve as responsibility-assignment guides for users of the framework. The developed frameworks and assignment guides are shown respectively in Figures 3.1, 3.2, and 3.3 and Tables 3.1, 3.2, 3.3, 3.4, and 3.5

## Figure 3.1 **Framework for the traditional system of procurement**

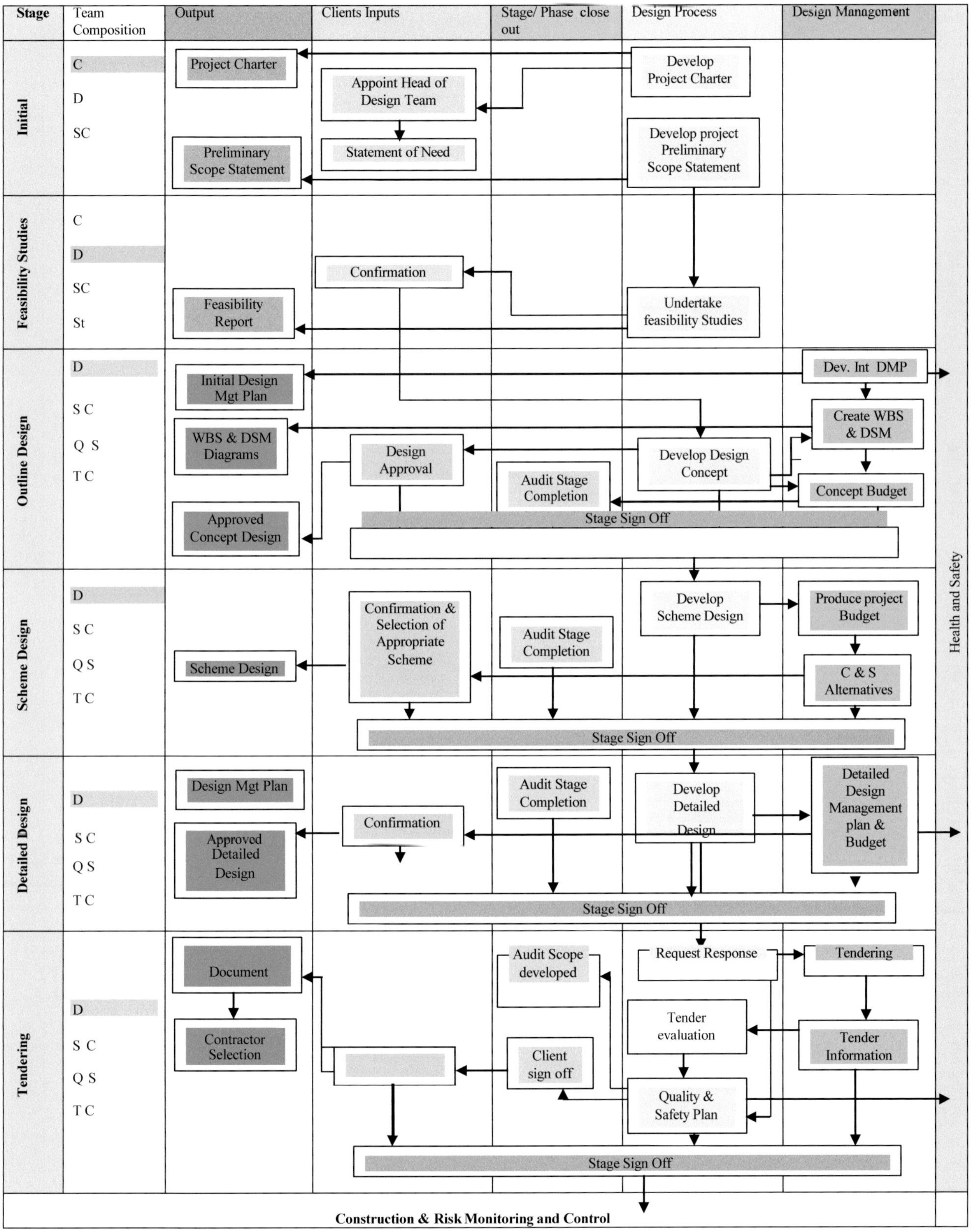

Stage | Team Composition | Output | Clients Inputs | Stage/ Phase close out | Design Process | Design Management

Initial
C
D
SC
Project Charter
Preliminary Scope Statement
Appoint Head of Design Team
Statement of Need
Develop Project Charter
Develop project Preliminary Scope Statement

Feasibility Studies
C
D
SC
St
Feasibility Report
Confirmation
Undertake feasibility Studies

Outline Design
D
S C
Q S
T C
Initial Design Mgt Plan
WBS & DSM Diagrams
Approved Concept Design
Design Approval
Audit Stage Completion
Develop Design Concept
Dev. Int DMP
Create WBS & DSM
Concept Budget
Stage Sign Off

Scheme Design
D
S C
Q S
T C
Scheme Design
Confirmation & Selection of Appropriate Scheme
Audit Stage Completion
Develop Scheme Design
Produce project Budget
C & S Alternatives
Stage Sign Off

Detailed Design
D
S C
Q S
T C
Design Mgt Plan
Approved Detailed Design
Confirmation
Audit Stage Completion
Develop Detailed Design
Detailed Design Management plan & Budget
Stage Sign Off

Tendering
D
S C
Q S
T C
Document
Contractor Selection
Audit Scope developed
Client sign off
Request Response
Tender evaluation
Quality & Safety Plan
Tendering
Tender Information
Stage Sign Off

Health and Safety

Construction & Risk Monitoring and Control

Figure 3.2 **Framework for the integrated system of procurement**

| Stage | Team | Output | Clients Inputs | Stage/ Phase close out | Design Process | Design Management | |
|---|---|---|---|---|---|---|---|
| Initial | C, P M, D, SC | Project Charter; Preliminary Scope Statement | Appoint PM; Appoint stakeholders; Statement of Need | | Develop Project Charter; Develop project Preliminary Scope Statement | | |
| Feasibility Studies | PM, C, SC, D, St | Feasibility Report | Confirmation | | Undertake feasibility Studies | | |
| | DT & C | Selected contractors | | | EXPRESSION OF INTEREST | | |
| Outline Design | PM, D, S C, Q S, T C | Initial Design Mgt Plan; WBS & DSM Diagrams; Approved Concept Design | Design Approval | Audit Stage Completion | Develop Scope; Develop Design Concept | Dev. Int DMP; Create WBS & DSM; Concept Budget | Health and Safety |
| | | | Stage Sign Off | | | | |
| Scheme Design | P M, D, S C, Q S, T C | Scheme Design | Confirmation of selected Appropriate Scheme | Audit Stage Completion | Develop Scheme Design | Produce project Budget; C & S Alternatives | |
| | | | Stage Sign Off | | | | |
| Detailed Design | P M, D, S C, Q S, T C | Design Mgt Plan; Approved Detailed Design | Confirmation | Audit Stage Completion | Develop Detailed Design | Detailed Design, Management Plan & Budget | |
| | | | Stage Sign Off | | | | |
| Tendering | PM, D, S C, Q S, T C | Tender Document; Contractor Selection | | Audit Scope developed; Client sign off | Request Response; Tender evaluation; Quality & Safety Plan | Tendering; Develop Tender Information | |
| | | | Stage Sign Off | | | | |

Construction Process & Risk Monitoring and Control

**Framework Key**

C – client

C & S – Create and select

D – designer

Dev. – develop

DMP – design management plan

DSM – dependency structure matrix

DT – design team

Mgmt. – management

PM – project manager

QS – quantity surveyor

SC – building services consultant

St – other stakeholders

TC – trades contractor

WBS – work breakdown structure

*Note:* Highlighted area in Team column depicts lead person at the respective stage.

Figure 3.3 **Design Management Framework for BSIW**

| Stage | Output | Clients Inputs | Stage/ Phase close out | Design Management by Building Services Consultants | Design Process Action | Design Management | Health and Safety |
|---|---|---|---|---|---|---|---|
| Initial | Design brief on services to feed project charter & preliminary scope statement | Statement of Need | | Receive Services brief<br>Define format for Design deliverables | Develop Project Charter<br>Develop project Preliminary Scope Statement | | |
| Feasibility Studies | Feasibility Report | Confirmation | | Provision and revision of data on BSIW | Undertake feasibility Studies | | |
| | | Feasibility Stage Sign Off | | | | | |
| Outline Design | Design Schedule & Proposal<br>Sketch & Schematic designs<br>Design Programme<br>Builders" Work & other design information | Design Approval | Audit Stage Completion | Create WBS & DSM to develop Schedule for design<br>Develop sketch & schematic design<br>Develop Cost, Risk & Quality outlines & analyses | Develop Scope<br>Develop Design Concept | Develop Initial Design Mgt Plan<br>Create WBS & DSM<br>Concept Budget | |
| | | | Outline Stage Sign Off | | | | |
| Scheme Design | Alternate designs<br>Cost & Schedule Plans<br>Detailed Schematic & technical drawings | Confirmation | Audit Stage Completion | Create, Review & Select alternative Designs<br>Evaluate Schemes using VM &QFD<br>Revise Cost, Risk, Quality & Schedule analyses<br>Develop detailed schematic and technical designs | Develop Scheme Design | Produce project Budget<br>Create & Select Alternatives | |
| | | | Scheme Stage Sign Off | | | | |

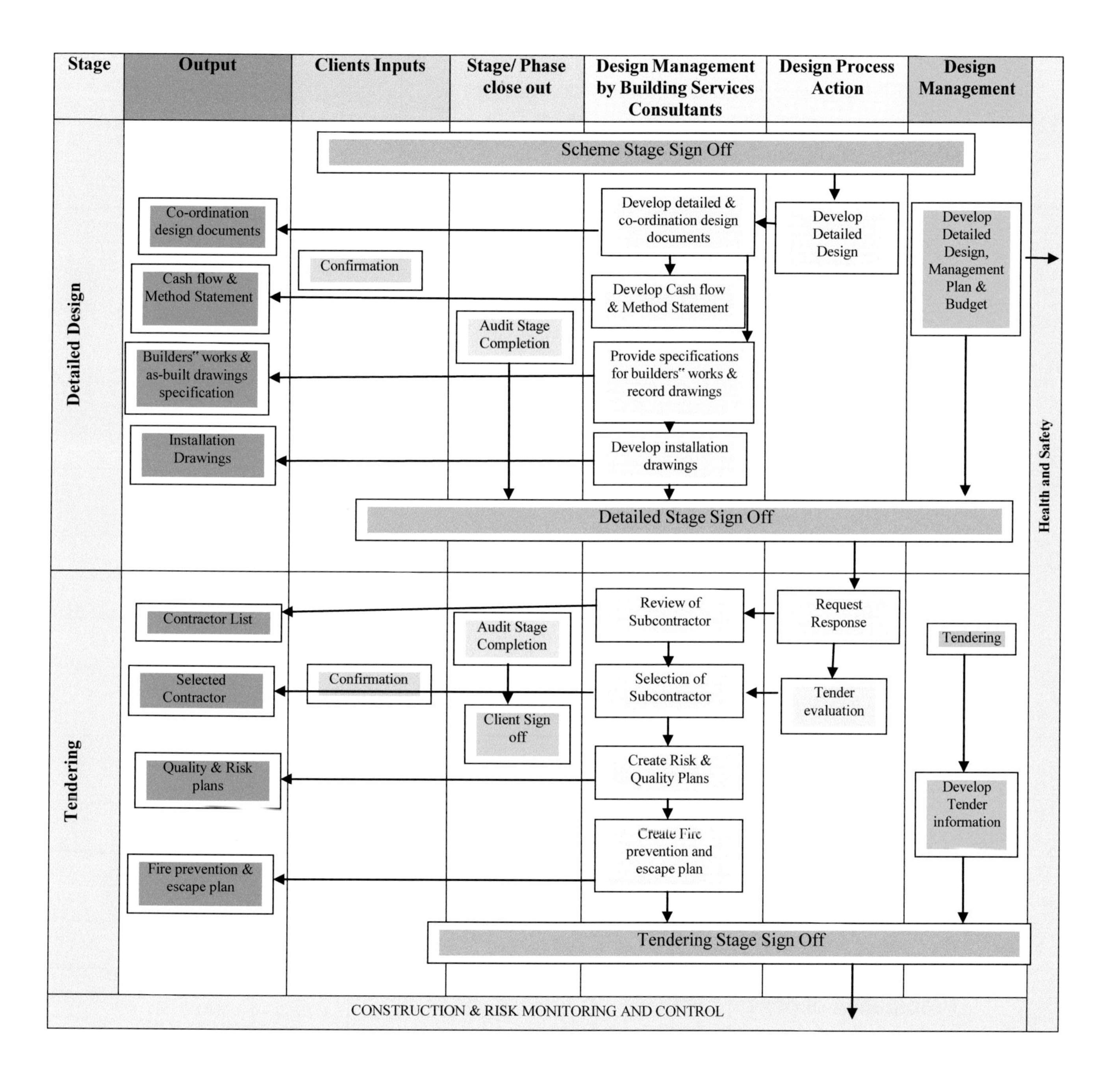
Stage
Output
Clients Inputs
Stage/ Phase close out
Design Management by Building Services Consultants
Design Process Action
Design Management
Detailed Design
Scheme Stage Sign Off
Co-ordination design documents
Develop detailed & co-ordination design documents
Develop Detailed Design
Develop Detailed Design, Management Plan & Budget
Confirmation
Cash flow & Method Statement
Develop Cash flow & Method Statement
Audit Stage Completion
Builders" works & as-built drawings specification
Provide specifications for builders" works & record drawings
Installation Drawings
Develop installation drawings
Detailed Stage Sign Off
Health and Safety
Tendering
Contractor List
Review of Subcontractor
Request Response
Audit Stage Completion
Tendering
Selected Contractor
Confirmation
Selection of Subcontractor
Tender evaluation
Client Sign off
Quality & Risk plans
Create Risk & Quality Plans
Develop Tender information
Fire prevention & escape plan
Create Fire prevention and escape plan
Tendering Stage Sign Off
CONSTRUCTION & RISK MONITORING AND CONTROL

**Deliverables of Building Services Installation Works at ach tage of the ramework**

Assignment Guide A, indicating allocation of design activity for Building Services Installation Works in stages 1 and 2

| Item no. | Design activity to carried out | | L | R | C | S | Remarks |
|---|---|---|---|---|---|---|---|
| 1 | Consulting appropriate authorities on matters concerned with the provision of building services works for the proposed site | | | | | | |
| 2 | Provision of physical data and environmental issues regarding building services works on site | | | | | | |
| 3 | Obtaining physical data for the provision of specific plants and design for: | Mechanical design | | | | | |
| | | Electrical design | | | | | |
| | | Public health design | | | | | |
| | | Fire engineering services design | | | | | |
| 4 | Provision of format for presentation of deliverables | Drawings | | | | | |
| | | Specifications | | | | | |
| 5 | Preparation of design brief | | | | | | |

**Table 3.1 Design activities for BSIW during initial and feasibility study stages**

**Key**: L – Lead; R – Review; C – Consult; S – Support

Assignment Guide B, indicating allocation of design activity for Building Services Installation Works in stage 3

| Item no. | Design activity to carried out | | L | R | C | S | Remarks |
|---|---|---|---|---|---|---|---|
| 1 | Preparation of schedule for design activities using WBS to break down packages and DSM to indicate dependencies | | | | | | |
| 2 | Preparation of risk and quality assessment for design activities | | | | | | |
| 3 | Informing other team members (structural and architectural) of sizes and weights for fittings, plants, and equipment | | | | | | |
| 4 | Establishment of builders' work principles for building services | | | | | | |
| 5 | Design proposal | Mechanical | | | | | |
| | | Electrical | | | | | |
| | | Public Health | | | | | |
| | | Fire services | | | | | |
| 6 | Preparation of sketch and schematic diagrams | | | | | | |
| 7 | Provision of design programme information | | | | | | |
| 8 | Early stage life cost outline and analysis | | | | | | |

**Table 3.2 Design activities for BSIW during the outline design stage**

Assignment Guide C, indicating allocation of design activity for Building Services Installation Works in stage 4

<table>
<tr><th>Item no.</th><th colspan="2">Design activity to carried out</th><th>L</th><th>R</th><th>C</th><th>S</th><th>Remarks</th></tr>
<tr><td>1</td><td colspan="2">Checking for adherence of compliance with regulations and requirements</td><td></td><td></td><td></td><td></td><td></td></tr>
<tr><td>2</td><td colspan="2">Creation of alternative designs, adequate calculations, planning, and advice to other team members on implications of design alternatives</td><td></td><td></td><td></td><td></td><td></td></tr>
<tr><td>3</td><td colspan="2">Evaluation of alternative schemes with the use of Value Management (VM) and Quality Function Deployment (QFD)</td><td></td><td></td><td></td><td></td><td></td></tr>
<tr><td>4</td><td colspan="2">Revision of risk and quality assessment</td><td></td><td></td><td></td><td></td><td></td></tr>
<tr><td>5</td><td colspan="2">Revision of costing and schedule activities</td><td></td><td></td><td></td><td></td><td></td></tr>
<tr><td rowspan="4">6</td><td rowspan="4">Review and selection of appropriate design</td><td>Mechanical</td><td></td><td></td><td></td><td></td><td></td></tr>
<tr><td>Electrical</td><td></td><td></td><td></td><td></td><td></td></tr>
<tr><td>Public Health</td><td></td><td></td><td></td><td></td><td></td></tr>
<tr><td>Fire services</td><td></td><td></td><td></td><td></td><td></td></tr>
<tr><td>7</td><td colspan="2">Preparation of detailed schematic and technical design drawings</td><td></td><td></td><td></td><td></td><td></td></tr>
</table>

**Table 3.3 Design activities for BSIW during the scheme design stage**

Assignment Guide D, indicating allocation of design activity for Building Services Installation Works in stage 5

| Item no. | Design activity to carried out | L | R | C | S | Remarks |
|---|---|---|---|---|---|---|
| 1 | Preparation of detailed design and coordination documents | | | | | |
| 2 | Creation of cash flow plan and method statement | | | | | |
| 3 | Provision of adequate builder's work information and workmanship specifications | | | | | |
| 4 | Define requirement for record/as-built drawings | | | | | |
| 5 | Preparation of installation drawings | | | | | |

**Table 3.4 Design activities for BSIW during detail design stage**

Assignment Guide E, indicating allocation of design activity for Building Services Installation Works in stage 6

| Item no. | Design activity to carried out | L | R | C | S | Remarks |
|---|---|---|---|---|---|---|
| 1 | Review of subcontractor information against design drawings | | | | | |
| 2 | Selection of appropriate trade's contractor for execution of the project | | | | | |
| 3 | Creation of risk and quality management plans | | | | | |
| 4 | Creation of fire prevention and escape plan for entire project | | | | | |

**Table 3.5 Design activities for BSIW during tendering stage**

## Design Management Framework Implementation Guide

*Outline*

The framework consists of six distinctive stages that run along with other processes to complete the framework. They include:

- **Team** – the composition of a team for each stage being considered (please note that they can be less than stated considering the type of project);
- **Outputs** – the derived outputs for the stage after all processes have been carried out;
- **Client inputs** – the requirements from the client or sponsor at the desired stage;
- **Design process** – activities carried out in the stage; and
- **Design Management** – management techniques employed during a particular stage.

Guidance matrices for the effective implementation of the framework have been outlined in Tables 3.1 to 3.5. These present what is required of the building services consultant at each stage. In each matrix, not all activities may be relevant for a particular project. In the case of such occurrence, the said activities can be eliminated for the purpose of the project being considered. For the assignment of Design Responsibilities, the parameters L, R, C, and S in a typical matrix are defined as:

- L – lead
- R – review
- C – consult
- S – support

In the completion of this matrix, a chart for assignment responsibility is created for the purpose of designing a typical project. Thus, various stakeholders responsible for leading, responsible for reviewing, and to be consulted, as well as those needed to support a particular BSIW activity, will be clearly defined.

**Stages in the Framework**

The stages include:

- **Initial stage** – This is the stage where the business need of the client is realized. A project charter is drawn up at this stage to appoint the project manager and other needed stakeholders for the design of the client's requirements. It also involves the creation of management plans and a budget for the project at hand. The output for this stage will be a project charter and a preliminary scope statement.
  - *A charter* is a document which identifies all stakeholders to be affected directly or indirectly by the project. It appoints the head of the design team (for the traditional system of procurement) or a project manager (for the management-oriented system of procurement and the integrated system of procurement). It formally authorizes the BSIW phase and documents the initial requirements to help satisfy stakeholders. The charter is carved out from the statement of work from the client (i.e. the client's brief) and also from the proficiency of experts.
  - *A preliminary scope statement/project brief* is a document which indicates the type of work to be carried out. This is developed after the client's requirements have been made known to the project design team. It helps translate the client's desire into achievable deliverables, bearing in mind the wants and the hows. The scope statement contains a plan for the management of time/schedule, cost, communication, quality, risk, and human resources.

These management plans created during this stage will help develop an initial budget and schedule outline for the project based on rule of thumb, analogous, parametric, or three-point estimation methods. This scope statement also identifies the internal and external factors that can affect the project and defines the work boundaries.

Team composition may include a services engineer, the client, the architect, and the project manager – where applicable – at this stage. The inclusion of the services engineer, making the team different from the likely project team composition in Ghana, can be attributed to the fact that building services take up an average of 35 per cent of the entire project cost and, when inappropriately managed, can lead to a variability a high as 129.68 per cent. With deliverable determinations being the prime responsibility of the services engineers and the architects, being present at the initial brief stage can reduce time commitment and conflict (on account of miscommunication and misinformation) that could occur along the line should the former be robed in the latter.

BSIW requirement at this stage: Physical data and environmental issues relating to the site can be identified for BSIW purposes.

**Feasibility studies stage** – During this stage, a study into why, how, and when the project is to be undertaken is carried out by a composed team for the client. The report from this study, the feasibility report, gives the sponsor the cost–benefit analysis of the proposed project and illustrates the need to carry out the project (if such information is necessary). External factors which may affect the project are all considered during this stage. The client/sponsor, after receiving this report, decides the need to carry out the project or decline its continuing. The output derived from this stage will be the feasibility study report.

Team composition at this stage may include a project manager, the client, a building services consultant, a designer/architect, and other stakeholder(s) needed to carry out an effective feasibility study.

BSIW requirement at this stage: Data gathering and studies carried out for preparation of the design brief. Appropriate regulatory and statutory permits acquired at this stage. It must be noted that in some projects, feasibility studies are carried out before the project is initiated, whereas others do it as part of the project procurement.

- **Outline design stage** – his is the stage at which the design brief of the client is developed into a concept for his or her approval. A design management plan for the entire BSIW is created at this stage, to help in the management of design throughout the project. A Work Breakdown Structure (WBS) is created at this stage. This in turn helps in the creation of a Dependency Structure Matrix (DSM) for appropriate scheduling and budgeting activities. A conceptual budget is created at this phase. The outputs of this outline stage are a design management plan, WBS and DSM diagrams, and an approved concept design.

o The *Work Breakdown Structure* is defined by PMI (2008) as "a deliverable-oriented hierarchical decomposition of the work to be executed by the project team to accomplish the project objectives and create the required deliverables." It is a tool used to define and group a project's discrete work elements (or tasks) in a way that helps organize and define the total work scope of the project.

- A WBS element may be a product, data, a service, or any combination of these. The WBS also provides the necessary framework for detailed cost estimating and control, along with providing guidance for schedule development and control. Additionally, the WBS is a dynamic tool and can be revised and updated as needed by the project manager or team leader.

o A *design structure matrix (DSM)* – also referred to as Dependency Structure Method, Dependency Structure Matrix, Problem- Solving Matrix (PSM), incidence matrix, N-square matrix, or design precedence matrix – is a compact matrix representation of a system or project. This approach is used to model complex systems in project planning and project management. Its analysis provides insight into how to manage complex systems or projects, highlighting information flows, task sequences, and iteration. It can help teams to streamline their processes based on the optimal flow of information between different interdependent activities.

A design structure matrix lists all constituent subsystems/activities and the corresponding information exchange and dependency patterns. Thus, it details what pieces of information are needed to start a particular activity, and shows where the information generated by that activity will lead. In this way, one can quickly recognize which other tasks are reliant upon information outputs generated by each activity.

This stage converts requirements into deliverables. The services engineer needs to be well-integrated so deliverables are favourably and adequately catered for.

The team composition at this stage may include the project manager, quantity surveyor, building services consultant, designer/architect, and trades contractor.

BSIW requirement at this stage: Sketch drawings and a sketch schematic are prepared for building services at this stage. At this stage, main service distribution routes can be identified, as well as the location and sizing of plant rooms, if applicable.

- **Scheme design** – This is the stage where the design scheme is developed. An initial scheme is developed and a schedule and budget report is created for this particular design. Alternative schemes created are reviewed using techniques of value engineering (VE) and quality function deployment (QFD) exercises. These schemes are also scheduled and budgeted for in the same way as the initial scheme. Once these schemes are compared,the most appropriate are selected for use at this stage by the client. The outputs from this stage are the alternative scheme designs, approved scheme with cost, and schedule estimates.

  The team composition at this stage may include the project manager, quantity surveyor, building services consultant, designer/ architect, and trades contractor

  BSIW requirement at this stage: Detailed schematic and technical design drawings can be prepared for the alternative schemes. At this stage, the spatial requirement as well as approximate sizing of pipes and ducts for building services works can be confirmed and determined respectively.

- **Detailed design** – The appropriate design that has been selected is now developed in detail at this stage into the works design for the project. The design management plan is also finalized at this stage for use in construction. This will constitute firm budget, schedule, and quality requirements, as well as a realistic risk register for the project to be carried out, for use of the project team and the client. Communication details – that is channels and means – are alsoclearly noted in the design management plan. The output of this stage is a detailed design and a design management plan.

The team composition at this stage may include the project manager, quantity surveyor, building services consultant, designer/architect, and trades contractor.

BSIW requirement at this stage: Detailed design and coordinated drawings, as well as builders' work details, are prepared at this stage.

- **Tendering** – This stage involves the selection of appropriate contractors to carry out the project. The output of this stage includes tender evaluation documents.

  The team composition at this stage may include the project manager, quantity surveyor, building services consultant, designer/architect, and trades contractor.

  BSIW requirement at this stage: Information received from contractors and subcontractors is reviewed. The installation drawing can thus be completed at this stage for construction to take place. It must, however, be noted that record or as-built drawings must be produced after construction to indicate any additions to and omissions from the installation drawings for future use.

## Operation of Created Framework

The developed framework with seven stages begins with the initial stage, where the client's need or desire is expressed. At this stage, the building services consultants, together with members of the building design team, assess the need for the project expressed by the client. A brief is received for building services at this stage. This will help feed the development of the entire project charter and scope statement. With the receipt of the brief, design deliverable formats are defined.

A feasibility study into issues regarding proposed BSIW is carried out at the next stage. During this exercise,the appropriate authorities are consulted on matters required for the provision of BSIW. Physical data needed to carry out the project are also sought at this stage. The output of this stage is a feasibility report.

After the feasibility report passes, and if there is a need to carry on with the proposed project, the scope of BSIW is developed at the outline stage. The developed scope is decomposed into packages using the work breakdown structure (WBS) technique. Dependencies, iteration, and information flow of work packages for schedule development are illustrated in a matrix with the aid of the dependency structure matrix (DSM). Sketches and sketch schematic drawings are produced during this stage. Cost, risk, and quality outlines are also developed at this stage. The stage outputs include:

i. schedule proposals for design;
ii. sketches and sketch schematic drawings; and
iii. design programme information for builders' works.

The stage signs off with an audit and moves to the scheme design stage.

At this stage, the main function of the services consultants is to create and review possible alternative design schemes for the selection of an appropriate one which best fits the client's requirements at an effective cost. The techniques of value management and quality function deployment (QFD) are employed at this stage. After a suitable scheme has been selected, cost, risk, quality, and schedule plans are reviewed. Detailed design and schematic drawings are produced at this stage. The stage output includes:

i. alternate design options;
ii. cost and schedule plans; and
iii. a detailed design and schematic drawings.

The stage signs off with an audit and moves into the detailed design stage.

At the detaildesign stage, the role of the building services consultant is to develop a detail design – one that is workable, has been selected from the created alternatives, and has the client's approval. Design is developed into detail and coordinated. In the context of design and build, cash flows and method statements are produced at this stage. Also, the detailed design specifications for builders' works as well as for the presentation of as-built or record drawings are created. Installation drawings for BSIW are developed at this stage. The output of the detail-design stage includes:

i. coordinated design documents;
ii. cash flow and method statements (where applicable);
iii. ibuilders' works and record drawings specifications; and
iv. installation drawings.

The stage is phased out with an audit.

The final stage considered in this framework is the tendering stage. The role of the building services consultants is to help with the selection of an appropriate subcontractor or trades contractor for the execution of the project. Tender and contract documents are produced, as are the drawings and specifications. These documents are responded to by sellers. The services consultant helps to review subcontract information against the design drawings during tendering.

An appropriate contractor is selected for the execution of BSIW after the tendering process. Together with this contractor, the services consultant completes the risk and quality plans, taking into consideration the organizational process assets of the contractor. A fire prevention and escape plan is created too. Stage deliverables include:

i. contractor list and selected contractor;
ii. quality and risk plans; and
iii. fire prevention and escape plan.

The stage phases out with an audit and with the client sign-off.

All plans, documents, drawings, and specifications obtained from these processes and stages outlined are channelled into the stages of construction and into the health and safety for the proposed project.

# SOURCES


Abdul-Rahman, H., Berawi, M. A., Berawi, A. R., Mohamed, O., Othman, M., and Yahya, I. A., "Delay Mitigation in the Malaysian Construction Industry", Journal of Construction Engineering and Management, 132/2 (2006), 125–33.

Adao I., "Quality Function Deployment – a Process for Continuous Improvement", Transactions from the Second Symposium on Quality Function Deployment, GOAL/QPC Research Committee, Michigan, 18–19 June 1989.

Ahmed, S. M., Sang, L. P., and Toria, M. Z., "Use of QFD in Civil Engineering Capital Project Planning", ASCE Journal of Construction, Engineering, and Management, 129/4 (2003), 358–68.

Akao Y., "Quality Function Deployment on Total Quality Management and Future Subject – QFD and TQM Series No. 1" (Japanese), Quality Control, 47/8 (1996), 55–64.

Akintoye, A., "Design and Build: A Survey of Construction Contractors' View", Construction Management and Economics, 12/2 (1994), 155–63.

Alarcon, L. F., and Mardones, D. A., "Improving the Design Construction Interface", Proceedings IGLC–6 (1998), Sao Paulo.

Allinson, K., Getting There by Design: An Architect's Guide to Design and Project Management (Oxford, Architectural Press, 1997).

Arain, F. M., and Low, S. P. "Measures for Minimizing Adverse Impact of Variations to Institutional Buildings in Singapore", Journal of Housing, Building and Planning, 10/1 (2003), 97–116.

Arditi, R. D., Akan, G. T., and Gurdamar, S., "Reasons for Delays in Public Projects in Turkey," Construction Management and Economics, 3 (1985), 171–81.

Armstrong, J., and Saville, A., Managing Your Building Services, Chartered Institution of Building Services Engineers (CIBSE) Knowledge Series (2002).

Ashworth, A., Cost Studies of Buildings (3rd edn, London, Longman, 1999).

Austin, S. A., Baldwin, A. N., and Newton, A. T., "Improved Building Design, Programming by Manipulating the Flow of Design Information", Construction Management and Economics, 12/5 (1994), 445–55.

—— "Simulating the Construction Design Process by Discrete Event Simulation", Proceedings of the International Conference of Engineering Design (ICED) 1995 (Prague, 1996), 762–72.

—— "A Data Flow Model to Plan and Manage the Building Design Process", Journal of Engineering Design, 7/1 (1996), 3–25.

—— "Techniques for the Management of Information Flow in Building Design", Proceedings of the International Conference on Information Technology in Civil and Structural Engineering (ITCSED) (1996).

Chartered Institute of Building, "Occasional Paper No. 40", 32.

—— Code of Practice for Project Management for Construction and Development (2nd edn, Hoboken, NJ, Blackwell Publishing, 2002).

Cohen L., Quality Function Deployment: How to Make QFD Work for You (Addison Wesley Longman Inc., 1995).

Coles, E. J., Design Management: A Study of Practice in the Building Industry (1990).

Coles, E. J., and Barritt, C. M. H., Planning and Monitoring Design Work (London, Pearson Educational Limited, 2000).

Connaughton, J. N., and Stuart D., Value Management in Construction (Construction Industry Research and Information Association, 1996).

Construction Industry Research and Information Association, Value Management in Construction: A Client's Guide, Construction Industry Research and Information Association Special Publication 129, 1996.

Creswell, J., Research Design: Qualitative and Quantitative Approach (Sage Publications, 1994).

Dansoh, A., "Strategic Planning Practice of Construction Firms in Ghana", Construction Management and Economics, 23/2 (2005), 163–68.

Dikmen, I., Birgonul, T. M., and Kiziltas, S., "Strategic Use of Quality Function Deployment (QFD) in the Construction Industry", Building and Environment, 2005, 245–55.

Edlin, N., and Hikle, V., "Pilot Study of QFD in Construction Projects", ASCE Journal of Construction, Engineering, and Management, 129/3 (2003), 314–29.

*Eppinger, S. D., Whitney, D. E., and Yassine, A. A., "The Design Structure Matrix", DSM <http://www.dsmweb.org/> accessed 31 Jan. 2007.*

*Faridi, A. S., and El-Sayegh, S. M., "Significant Factors Causing Delay in the UAE Construction Industry", Construction Management and Economics, 24 (2006), 1167–76.*

*Fisher, C., Researching and Writing a Dissertation: A Guide for Business Students (London, Pearson Education Limited, 2007).*

*Frazer, L., and Lawley M., Questionnaire Design and Administration: A Practical Guide (Brisbane, John Wiley & Sons, 2000).*

*Frimpong, Y., Oluwoye, J., and Crawford, L., "Causes of Delay and Cost Overruns in Construction of Groundwater Projects in Developing Countries: Ghana as a Case Study", International Journal of Project Management, 21/5 (2003), 321–26.*

*Generic Design and Construction Process Protocol, EPSCR, IMI.*

*Ghana: Demography of Ghana, Statistical Survey Report, 1998.*

*Gray, C., and Hughes, W., Building Design Management (2nd edn, Oxford, Butterworth-Heinemann, 2000).*

*—— Building Design Management (3rd edn, Oxford, Butterworth-Heinemann, 2006).*

*Gray, C., Hughes, W., and Bennette, J., The Successful Management of Design (Reading, Centre for Strategic Studies in Construction, 1994).*

*Great Britain. British Standard: Design Management Systems – Part 10: Vocabulary of Terms Used in Design Management, BS 7000 (March 2008).*

*—— Design Management Systems – Part 4. Guide to Managing Design in Construction, BS 7000 (March 1996).*

*Greeno, R., and Hall. F, Building Services Handbook: Incorporating Current Building and Construction Regulations (2nd edn, Oxford, Butterworth-Heinemann, 2003).*

*Gregson, J., "Rethinking Construction", report from Construction Task Force (London, Department of the Environment, Transport, and the Regions, 1998).*

*Griffin, A., and Hauser, J., "The Voice of the Customer: Technical Report Working Paper" (Cambridge, MA, Marketing Science Institute, 1992), 92–106.*

*Hall, F., Building Services Equipment (Oxford, Elsevier Butterworth-Heinemann, 2001).*

*Hall, F., and Greeno, R., Building Services Handbook (5th edn, Oxford, Butterworth-Heinemann, 2009).*

*Hauser, J. R., "How Puritan-Bennett Used the House of Quality", Sloan Management Review, 34/3 (1993), 61–70.*

*Hauser, J. R., and Clausing, D., "The House of Quality", Harvard Business Review, 32/5 (1998), 63–73.*

*Hayden, G. W., and Parsloe, C. J., Value Engineering for Building Services, Building Services Research and Information Association Application Guide 15/96 (Borune Press, 1996).*

*Holbrough, M., Building Services Engineer – Job Description and Activities (London South Bank University, AGCAS, and Graduate Prospect Limited, 2009).*

*Hyatt, P., In Practice (London, Emap Construct, 2000).*

*Kelly, J., Male, S., and Graham, D., Value Management of Construction Projects (Oxford, Blackwell Science, 2004).*

*King, R., "Listening to the Voice of the Customer: Using the Quality Function Deployment System", National Productivity Review, 1987, 277–81.*

*Kish, L., Survey Sampling (New York, Wiley & Sons Inc., 1965).*

*Knight, A., Griffith, A., and King, A. P., "Supply Side Short Circuiting in Design and Build Projects", Management Decision, 40/7 (2002), 655–62.*

*Knight, J., and Jones, P., Building Service (Amsterdam, Elsevier Butterworth-Heinemann, 1995).*

*Koskela, L., Ballard, G., and Tanhuanpaa, V., Towards Lean Design Management, Proceedings of the IGLC–5 (Australia, 1997).*

*Lai-Kow, C., and Ming-Lu, W., "A Systematic Approach to Quality Function Deployment with a Full Illustration Example", International Journal of Management Science, 33/2 (2005), 119–39.*

*Langmaid, J., Choosing Building Services: A Practical Guide to System Selection: A Building Services Research and Information Association Guide, BG 9/24 (September 2004).*

*Lincoln, Y. S., and Guba, E. G., Naturalistic Inquiry (Beverly Hills, Sage Publications, 1985).*

*Long, N. D., Ogunlana, S., Quang, T., and Lam, K. C., "Large Construction Projects in Developing Countries: A Case Study from Vietnam", International Journal of Project Management, 20/7 (2004), 553–61.*

Low, S. P., and Yeap, L., "QFD in Design and Build Projects", Journal of Architectural Engineering, 7/2 (2001), 30–39.

Mallon, J. C., and Mulligan, D. E., "QFD – A System for Meeting Customer Needs", ASCE Journal of Construction, Engineering, and Management, 119/3 (1993), 516–31.

Manoharan, R., "Subcontractor Selection Method Using Analytic Hierarchy Process", unpublished Master's thesis, University of Technology, Malaysia, 2005.

Marrano, S. J., and DiLouie, C., Electrical System Design and Specification Handbook for Industrial Facilities (Fairmount Press, 1996).

Masterman, J. W. E., Introduction to Building Procurement Systems (London, E. & F. N. Spon, 2002).

Mulcahy, S., "Building Services", The CIBSE Journal, 1979.

Nani, G., "Conceptual Framework for Development of a Standard Method of Measurement of Building Works for Ghanaian", unpublished, PhD thesis report, Kwame Nkrumah University of Science and Technology, Ghana, 2009.

Naoum, S. G., Dissertation Research and Writing for Construction Students (Amsterdam, Elsevier, 2007).

Odeh, A. M., and Battaineh, H. T., "Causes of Construction Delay: Traditional Contracts", International Journal of Project Management, 20 (2002), 67–73.

Oloufa, A., Hosni, Y., Fayez, M., and Axelsson, P., Using DSM for Modeling Information Flow in Construction Design Projects (Taylor & Francis Ltd, Spon Press, 2004).

Oppong, B., "Construction Delays in Ghana", unpublished MSc thesis report, Kwame Nkrumah University of Science and Technology, Ghana, 2003.

Osei-Tutu, E., "A Rational Procedure for the Selection of Appropriate Procurement Systems in Ghana", unpublished MSc thesis report, Kwame Nkrumah University of Science and Technology, Ghana.

Patrascu, A., Construction Cost Engineering Handbook (New York, Marcel Dekker Inc., Kindle edition, 1998).

Potter, M., Planning to Build: A Practical Introduction to the Construction Process, Special Publication No. 113 (London, Construction Industry Research and Information Association, 1995).

Prasad, B., Con-current Function Deployment: An Emerging Alternative to QFD, Conceptual Framework –Advantage in Concurrent Engineering, ed. Con, Sobolewski, and M. Fox (Lancaster, Technomic Publishing, 1996).

Project Management Institute, A Guide to Project Management Body of Knowledge (PMBOK® Guide) (1st edn, Project Management Institute Inc., 1996).

—— A Guide to Project Management Body of Knowledge (PMBOK® Guide) (4th edn, Project Management Institute Inc., 2008).

Rao, V., and Katz, R., "Alternative Multidimensional Scaling Methods for Large Stimulus Sets", Journal of Marketing Research, 8 (1992), 488–94.

RIBA, Outline Plan of Work RIBA (London, Royal Institution British Architects Publications, 2007).

Sambasivan, M., and Soon, Y. W., "Causes and Effects of Delays in Malaysian Construction Industry", International Journal of Project Management, 25 (2007), 517–26.

Saunders, M., Lewis, P., and Thornhill, A., Research Methods for Business Students (London, Pearson Education, 2007).

Schrage, M., No More Teams: Mastering the Dynamics of Creative Collaboration (New York, Doubleday, 1990).

Seeley, I., Quantity Surveying Practice (2nd edn, London, Macmillan Press Ltd, 1997).

Sengupta, B., and Guha, H., Construction Management and Planning (India, Tata McGraw Hill Publishing Company Limited, 1995).

Serpell, A., and Wagner, R., "Application of Quality Function Deployment (QFD) to the Determination of the Design Characteristics of Building Apartments", in L. A. Larcon (ed.), Lean Construction (Rotterdam, The Netherlands, Balkema, 1997), 355–63.

Sheridan, J. C., and Lyndall G. S., SPSS Analysis without Anguish 10.0 for Windows (Brisbane: John Wiley & Sons, 2001).

Stehn, L., and Bergstrom, M., "Integrated Design and Production of Multi-Storey Timber Frame Houses Effects Caused by Customer Oriented Design", International Journal of Production Economics, 77/3 (2002), 259–69.

Steward, D., The Design Structure Systems, General Electric Report No. 67APE6 (San Jose, 1967).

—— Planning and Managing the Design of Systems, Proceedings of Portland International Conference on Management of Engineering and Technology (Portland, OR, 1991), 27–31.

*Temponi, C., Yen, J., and Tiao, W. A., "Theory and Methodology House of Quality: A Fuzzy Logic-Based Requirements Analysis", European Journal of Operational Research, 117 (1999), 340–54.*

*Wild, J., Site Management of Building Services Contractors (London, E. & F. N. Spon, 1997).*

*Yang, Q. W., Mohammad, D., and Sui, P. L., "A Fuzzy Quality Function Deployment System for Buildable Design Decision Makings", Automation in Construction, 12/4 (2003), 381–93.*

*Yasin, J., "An Investigation into the True/Full Cost of Building Services", unpublished MSc thesis, Heriot Watt University, 40-7315, 2004.*

*Yin, R. K., Case Study Research: Design and Methods (3rd edn, London, Sage Publications, 2003).*

*Zairi M., and Youssef, M., "Quality Function Deployment: A Main Pillar for Successful Total Quality Management and Product Development", International Journal of Quality and Reliability Management, 12/6 (1995), 9–23.*

*Zikmund, W., Business Research Methods (Dryden Press, 1997).*

*Zucchelli, F., "Total Quality and QFD", in 1st European Conference on Quality Function Deployment (Galgano & Associates, 1992), 25–26.*